A B O V E &

B E Y O N D

MIR SPACE STATION

Published by Smart Apple Media

123 South Broad Street

Mankato, Minnesota 56001

Copyright © 2000 Smart Apple Media.

International copyright reserved in all countries. No part

of this book may be reproduced in any form without written

permission from the publisher.

Printed in the United States of America.

Photos: nasa/kennedy space center; page 29—corbis

Design and Production: EvansDay Design

Library of Congress Cataloging-in-Publication Data

Bernards, Neal, 1963–

Mir space station / by Neal Bernards

p. cm. — (Above and beyond)

Includes index.

Summary: Describes the Mir space station, its background,

construction, problems, and contributions to our

knowledge and experiences in space.

ISBN 1-58340-049-4

1. Mir (Space Station)—Juvenile literature. [1. Mir (Space Station).]

I. Title. II. Series: Above and beyond (Mankato, Minn.)

TL789.8.R92M573   1999

629.44'2'0947—DC21   98-18246

3 5 7 9 8 6 4 2

MIR SPACE STATION

NEAL BERNARDS

ABOVE & BEYOND

MIR'S COOLING FANS and electronic instruments hummed on an even tone as the **space station** drifted in **orbit** above Earth ✳ An unmanned Progress cargo vessel approached with its load of needed supplies ✳ Suddenly, shouts shattered the calm inside *Mir* ✳ The Progress was out of control and coming in too fast ✳ The Russian commander frantically tried to regain control of the wayward vessel, but all attempts failed ✳ An American astronaut scrambled to prepare the *Soyuz* descent vehicle—the crew's only lifeboat home ✳ A dull thud shook the station, and a shrill warning bell sounded ✳ The ship had crashed into *Mir* ✳

A **space station** *is a spacecraft designed to carry a crew in orbit for long periods of time.*

To **orbit** *is to revolve around a moon or planet.*

Salyuts, Shuttles, and *Skylab*

The Soviet Union launched the core **module** of *Mir* on February 20, 1986. Since September 5, 1989, there have always been at least two people living in orbit aboard the space station. *Mir* has served as a laboratory where scientists can study the effects of life in space on the human body. Its crew members have experienced triumphs, near tragedies, and the special demands of everyday life in a weightless environment.

Russians built the space station *Mir* in an attempt to leap ahead of the United States in this area of space technology. This competition—known as the space race—between the Soviet Union and the U.S. began in the 1950s. Both nations wanted to be the first to launch a satellite, the first to put a man in space, and the first to reach the moon.

In the 1950s and 1960s, the Soviet Union was clearly the leader. The Soviets launched the very first spacecraft, a small satellite called *Sputnik 1,* in October 1957. This successful launch shocked Americans, many of whom had long assumed that the United States was the world leader in every field, including space technology. *Sputnik 1* put this belief in serious doubt. The Soviets had clearly won the first round of the space race.

The Soviets also won the second round. On April 12, 1961, they sent **cosmonaut** Yuri Gagarin into space aboard *Vostok 1*. Gagarin made one 108-minute orbit of Earth, then parachuted to the ground from two miles up.

Just three weeks later, on May 5, 1961, the U.S. launched astronaut Alan Shepard into space aboard a **capsule** called *Freedom 7*. Although Shepard was in space for only about 15 minutes, the flight was an important step toward America's goal of landing a man on the moon.

*A **module** is a part of a spacecraft that can be detached and operated separately.*

Cosmonaut is the name given to astronauts from the former Soviet Union.

The first American astronauts experienced space in capsules.

*A **capsule** is the part of a spacecraft in which people travel.*

Eight years passed before the *Apollo 11* **mission** achieved that goal. On July 20, 1969, U.S. astronauts Neil Armstrong and Edwin "Buzz" Aldrin boarded *Apollo 11*'s lunar module and descended to the lunar surface, becoming the first humans to set foot on the moon.

After the Americans beat them to the moon, the Soviets turned their attention to space stations. *Salyut 1* through 7 gave Soviet cosmonauts valuable experience living in space. Each *Salyut* mission lasted longer than the previous one; *Salyut 6* and *7* represented a "second generation" of space stations that stayed in space for years rather than months.

A **mission** *is a program or plan to achieve specific goals.*

In the U.S., the National Aeronautics and Space Administration (NASA) rushed to catch up with the Soviets and their space station technology. NASA hastily built *Skylab*, the first American space station, out of leftover

Skylab, *the United States' first space station.*

parts from *Apollo* and *Saturn* rockets. *Skylab* was launched in 1973.

The 118-foot American space station was larger than any of the *Salyuts*, but it was plagued with problems from the beginning. The temperature inside *Skylab* was about 110 degrees Fahrenheit (43 degrees C)—too warm for human comfort. The space station's shower leaked, and important pieces of equipment needed constant repair. After astronauts visited *Skylab* only three times, NASA sealed it shut in 1974. The abandoned space station drifted for five years before falling to Earth on July 11, 1979.

America was the first and only nation to land men on the moon.

Cooperation in Space

Despite their competing space programs, the Soviet Union and the U.S. began a cooperative mission on July 17, 1975. This milestone in space technology was the result of an agreement between U.S. President Richard Nixon and Soviet Premier Aleksey Kosygin.

During the mission, the American *Apollo 18* **docked** with the Soviet *Soyuz 19*. For two days, the astronauts and cosmonauts shared meals, carried out scientific experiments together, and visited one another's spacecraft. This cooperation was particularly amazing because it came at the peak of the Cold War, a period when the Soviet Union and the U.S. were sworn enemies.

Russian–American cooperation in space would not happen again for nearly 20 years. In 1993, the U.S. agreed to pay Russia $473 million for four years of time aboard *Mir*. For this price, American astronauts were able to train with cosmonauts in Star City, 30 miles outside of Moscow, and work alongside them on *Mir*. For an amount that was less than one-tenth of NASA's yearly budget, U.S. astronauts would gain the experience of living and working in the world's only functioning space station.

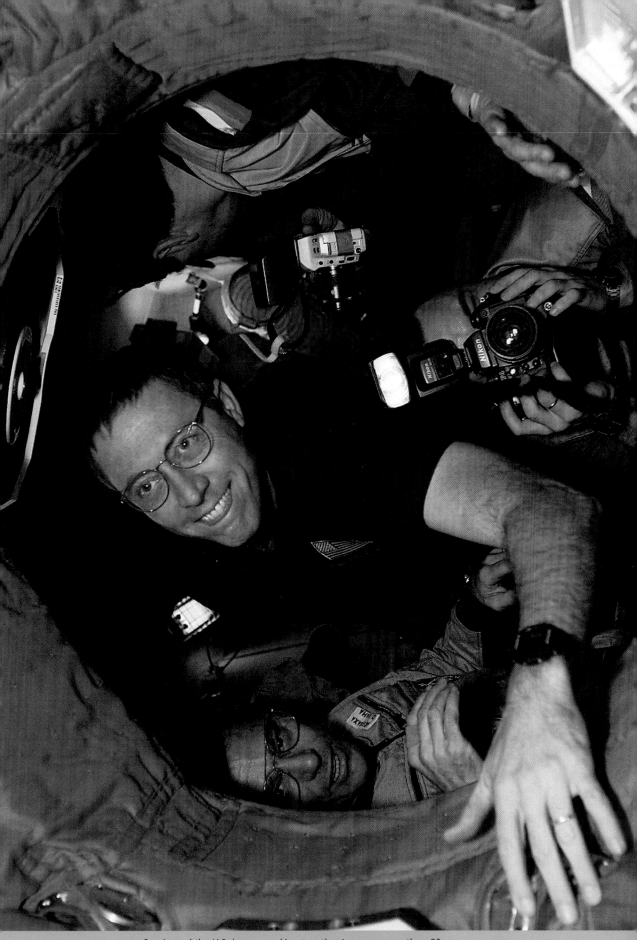

Russia and the U.S. began working together in space more than 23 years ago.

Docking *is the physical connection of two spacecraft in space.*

Critics of the *Mir* agreement thought that the U.S. should not give money to its former rival, arguing that NASA could build a better space station without Russian help. Supporters, on the other hand, believed that cooperating with the Russians was a good political move. Following the 1991 collapse of the Soviet Union, many U.S. leaders wanted to show support for the reform occurring in Russia. By supporting Russia's space technology program, the U.S. would help the Russian economy and strengthen peaceful ties between the two countries.

Teamwork among Mir's crew has been an essential part of the space station's success.

Although NASA has enough funding to develop its own manned space program, the idea has received little support. In the 1960s, millions of Americans supported NASA's goal of reaching the moon. But the 1986 *Challenger* space shuttle explosion—which killed all seven astronauts on board—made many Americans question the need for risky manned space flights.

Many people believe that sending unmanned space probes to study distant planets is currently the safest and wisest approach to the study of space. NASA officials decided that the *Mir* agreement might be the only way to keep the American program of manned space flight alive.

Docking Mir *with space shuttles takes patience and precision.*

Life aboard *Mir*

In March 1995, astronaut Norman Thagard became the first American to live on *Mir*. Before joining the Russian crew in space, Thagard trained with cosmonauts for a year in Star City; by the time he arrived on *Mir*, Thagard spoke Russian and was well-prepared.

At first glance, *Mir* looks like a Tinker Toy collage of bits and pieces. The first portion in space was the *Mir* core module, launched in 1986. This module, which replaced the Soviets' *Salyut 7* space station, contains the space station's primary living space, a kitchen, private crew cabins, and an exercise bike.

Over the next 10 years, scientists added six modules to the core module, making *Mir* the largest orbiting space vehicle ever built. The first attachment was the *Kvant 1* scientific module, added in 1987. This module contains instruments for changing *Mir*'s position in space and controlling temperature, as well as equipment needed to perform experiments in astrophysics and biotechnology.

Kvant 2 joined *Mir* in 1989. It too is a scientific module, but it also contains life support equipment, a bathroom,

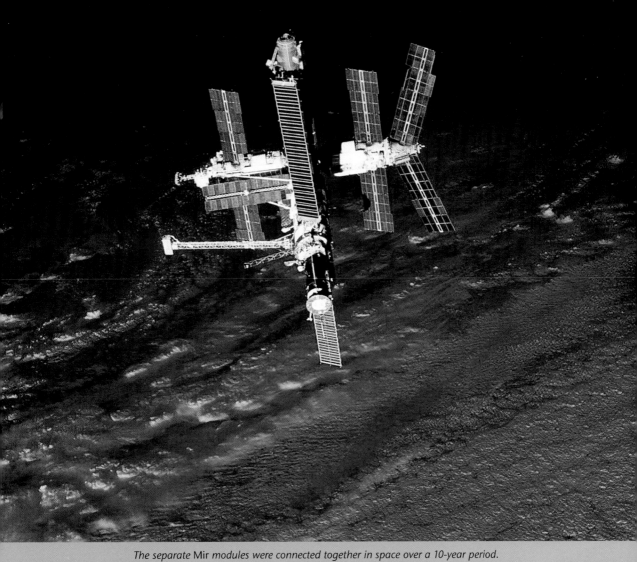

The separate Mir modules were connected together in space over a 10-year period.

solar panels, and an air lock that allows crew members to take **space walks**.

The Russians added *Krystall* to *Mir* in 1990. This module contains equipment needed for research on producing biological and chemical materials, as well as a docking port that allows other spacecraft to dock with *Mir.*

In 1995, the Russians added another module, this one called *Spektr*. This component of the space station carries solar panels and houses equipment used to study the effects of space on the human body. The most recent addition to *Mir* was the *Priroda* module, which arrived in 1996. This module carries additional equipment used to research human biology.

All of the modules serve different purposes, but they all work together to give the space station its power, direc-

Solar panels *are solar cells that produce energy from sunlight.*

Solar panels power much of Mir's *equipment.*

tion, and stability. *Mir* was originally scheduled to spend just five years in space, but its solid design has allowed it to remain in space far longer.

Russian designers worked hard to create a pleasant living space inside *Mir*. The carpet is dark green, the walls are light green, and the ceiling is white. This color layout helps crew members maintain a sense of direction, which is not always easy to do in space.

Living quarters on *Mir* are separate from work stations. This separation began on *Salyut*, whose cosmonauts felt it

A **space walk** *is a period of activity outside of an orbiting spacecraft.*

Mir crew members Gennady Strekalov and Greg Harbaugh working on the station's utilities.

was important for men and women in space to have their own personal areas.

Perhaps the most extreme difference between life on Earth and life on *Mir* is the gravity—on *Mir,* there is none. Crew members float through the space station as if they were swimming underwater. They can push themselves from one end of *Mir* to the other with little effort. This weightlessness can be fun, but it takes some getting used to. Anything that is not strapped down tends to drift about the cabin. As a result, the crew must secure everything on board.

Weightlessness poses a problem for sleep, too. It is difficult to lie down in space, because there is no gravity to create a "down" direction. To solve this problem, crew members anchor their sleeping bags to the walls. When they want to sleep, they use Velcro to strap themselves into their bags for the night so they won't float around the space station.

On Earth, people's muscles work against gravity every time they stand, walk, or climb a flight of stairs. Weightlessness, however, can cause muscles to go weak through disuse. During early space missions, astronauts and cosmonauts would return to Earth so weak that they would

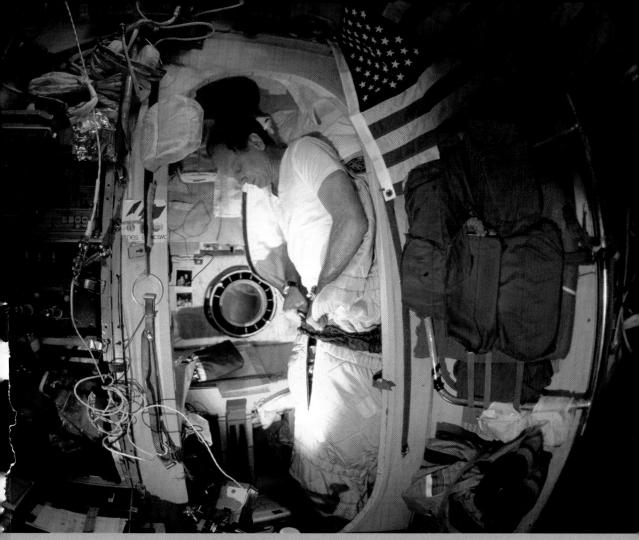

American astronaut Norman Thagard prepares for a nap on Mir.

be unable to stand. Eventually, space travelers learned that they had to exercise daily to stay fit. Crew members on *Mir* spend up to two hours a day exercising on a bike and stretching elastic bands.

Bathing presents a special challenge on *Mir*. Water does not fall in weightless conditions, so Russian engineers designed a special shower using suction and vacuums. But this shower system broke down long ago; crew members can now only wipe themselves clean with a wet cloth. Valery Polyakov, who spent a record 438 days on board *Mir,* was not able to take a single shower.

Just like people on Earth, *Mir* crew members generate a certain amount of waste and garbage. To dispose of these items, Russian scientists developed a unique garbage disposal system. Anything that cannot be reused is loaded

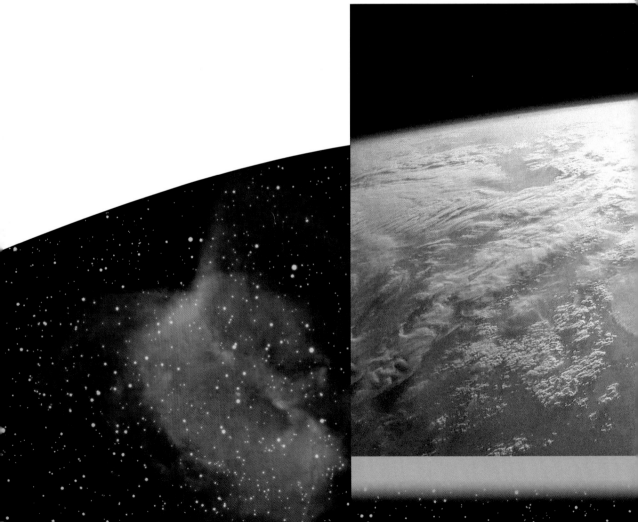

into an unmanned Progress cargo ship, which is then sent into space. After a short time, the trash-carrying ship loses altitude and burns up in the earth's atmosphere.

Mir crew members perform a variety of jobs. They pilot the space station and monitor *Mir*'s instruments to make sure the station's solar panels—which power *Mir*—are pointed toward the sun. Crew members must also ensure that the spacecraft is producing breathable oxygen and that it has adequate fuel supplies.

Those living on Mir *have a constantly spectacular view of Earth.*

Mir's crew maintains daily radio contact with flight control in the Baikonur **cosmodrome**: the Russian space center. Flight controllers there keep *Mir* on a schedule. They tell crew members when to fire their rockets to stay in proper orbit, when supply vessels will arrive, and when the next space shuttle or *Soyuz* crew will come to relieve them.

*A **cosmodrome** is a Russian space center.*

When other spacecraft dock with *Mir*, crew members must pilot the vessels into the docking port. This operation requires great precision; being off by as little as four inches when the two spacecraft meet can lead to disaster. The space station is equipped with a computer guidance system for docking operations, but the system has broken down in the past. When this happens, crew members have to steer the approaching craft manually.

Mir crew members carry out scientific experiments—in astronomy, biology, physics, and other fields—in the

An Atlantis *astronaut photographs the orbiting station.*

Spektr and *Kvant* modules. Crew members may study the stars, the growth of plants in space, or the effects of space travel on humans. When an experiment is finished, all of the data and materials are packed aboard a *Soyuz* and returned to scientists on Earth.

Despite living on a ship full of high-tech gadgets and scientific equipment, most crew members say that their favorite activity is looking out of *Mir*'s porthole-sized windows. Because *Mir* orbits Earth at 17,500 miles (27,540 km) per hour, its crew can enjoy as many as 16 sunrises and 16 sunsets each day.

An image from Mir of the docked Atlantis with the earth far below.

Famous for the Wrong Reasons

Despite its many successes and its overall durability, *Mir* has become famous for its numerous problems. The space station has suffered just about every imaginable difficulty short of destruction.

Mir's problems started before Americans arrived on board the space station. In early 1994, a Progress cargo vessel docked with *Mir*, bringing badly needed food, oxygen, and water to the crew of three orbiting cosmonauts. When the cosmonauts opened the hatch, however, they discovered that most of the food was gone. Russian ground-support workers—who were not making enough money to feed their families—had stolen it prior to launch.

The cosmonauts soon became so hungry that they ate leftovers from previous missions, some of which were two years past their expiration date. They also had to drink water recycled from their own breath and sweat. A second Progress cargo ship with new supplies was quickly sent, but the ground controllers directing the ship missed *Mir* on their first two docking attempts. There was only enough fuel for one more attempt; if it failed, the cosmo-

Atlantis's *shuttle pilot's view of the Russian space station.*

nauts would have to return home, leaving *Mir* unmanned for the first time in five years. At the last moment, cosmonaut Yuri Malenchenko lined *Mir* up perfectly for a **rendezvous** with the cargo ship.

Mir survived a year of mistakes, mishaps, and accidents in 1997. In February, oxygen-producing candles started a fire in one of *Mir*'s labs. Crew members were using the

A **rendezvous** *is a meeting between two spacecraft.*

JUNE 29, 1995

The space shuttle Atlantis *makes its first docking with* Mir.

candles because *Mir*'s main oxygen-generating system had broken down. Cosmonauts managed to put the fire out with an extinguisher, but the blaze caused extensive smoke damage to *Mir*'s interior.

In April 1997, *Mir*'s cooling system leaked, causing temperatures to soar inside the space station. The air purification system then failed, leaving crew members short of breath. The problems were again solved, but not before the crew had endured considerable discomfort.

Mir's worst crisis began with a routine rendezvous between the space station and another Progress cargo ship. On June 25, 1997, *Mir* commander Vasiliy Tsibliev, who was piloting the cargo ship by remote control, lost control of the cargo vessel during the docking operation. It raced past the docking area and struck *Spektr* and its solar panels. As *Spektr* began to leak oxygen, the crew had to shut the damaged module down entirely, losing half of *Mir*'s power supply in the process. The crew prepared to abandon *Mir* in the *Soyuz* lifeboat. After 48 tense hours of

Mir's *June 25, 1997,* collision with a Progress cargo ship left one of its solar panels damaged.

A radiator panel also took damage in the crash.

repairs, however, crew members finally stabilized *Mir* and completed the docking.

Mir lost power again on July 16, 1997, when a cosmonaut pulled the wrong cable during repairs from the earlier crash. Just three weeks later, on August 5, both oxygen generators failed. To breathe, the crew had to use oxygen-producing candles like those that had started the earlier fire. Then, on September 10, *Mir*'s computers failed for a third time. The space station lost all power.

Many people thought that the aging station would not survive 1997. Any one of the problems it encountered could have been fatal. But, as supporters of *Mir* point out, the fatal accident did not occur. In fact, the experience gained by building and operating *Mir* will undoubtedly lead to safer space travel in the future.

Mir's mission control center in Russia.

The experience gained from Mir will lead to bigger and better space vehicles in the future.

The Future of
Space Stations

Decades of rocket launches, research, and experience in space went into the creation of *Mir*. The crowning achievement in a long line of successful Soviet space stations, *Mir* has fostered a rewarding partnership between the Russian and American space programs.

Mir's success has inspired other cooperative efforts in space. Sixteen nations will build components for a spacecraft called the International Space Station. Its design will combine Russia's experience in long-term space flight with America's cutting-edge technology. Scientists hope to use the space station as a jumping-off point for future missions, including a two-year manned flight to Mars. The station's first two modules—*Unity* and *Zarya*—were joined in space on December 6, 1998.

Most scientists agree that space exploration can move forward only with cooperation between nations, particularly between Russia and the U.S. No one knows for certain when the International Space Station will be completed, but the crew of *Mir* will remain in space until the new station is up and running. Abandoning *Mir* would seem to many like taking a big step backward. After all, *Mir* has made space into a second home.

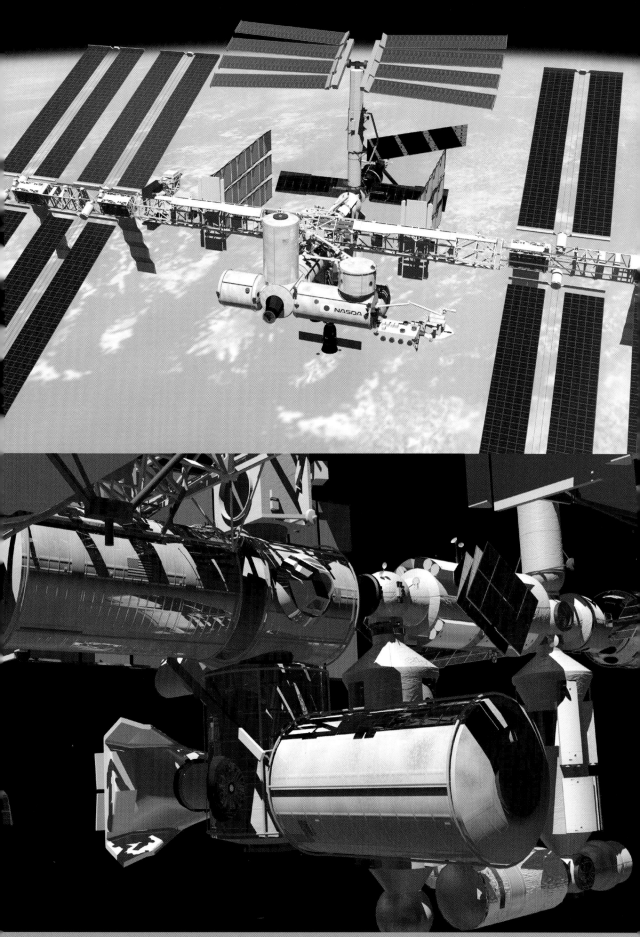

Assembly of the International Space Station began in December 1998 and will continue in the years ahead.

INDEX